YOUR KNOWLEDGE HAS VALUE

- We will publish your bachelor's and master's thesis, essays and papers

- Your own eBook and book - sold worldwide in all relevant shops

- Earn money with each sale

Upload your text at www.GRIN.com and publish for free

Understanding Contract Termination Dynamics in Public Building Projects

Anteneh Birhanu

Bibliographic information published by the German National Library:

The German National Library lists this publication in the National Bibliography; detailed bibliographic data are available on the Internet at http://dnb.dnb.de.

ISBN: 9783346886651

This book is also available as an ebook.

Hawassa University Institute of Technology

Faculty of Civil Engineering Built Environment

Department Of Construction Technology and Management (Pg.)

MSc thesis

Project Proposal

Title: - Study on Public Construction Projects Contract Termination and Its Consequences

Prepared BY

Anteneh Birhanu

Hawassa, Ethiopia

May 9 2023

List of Abbreviations

- SNNPRS _________________Southern Nations Nationalities Regional State
- IPDC___________________Industrial Parks Development Corporations
- PCPCT__________________Public Construction Project Contract Termination
- PIBIP _________________ Public industrial Building and Infrastructure projects
- SME _________________Small scale and Micro enterprises

Summery:

This research aims to investigate the causes and effects of construction contract termination in public building projects owned by the SNNPRS Industrial Parks Development Corporation. Titled "Study on Public Construction Projects Contract Termination and Its Consequences," the study will delve into understanding the rate of contract termination, examining the underlying causes, and evaluating the impacts of termination on social and economic aspects. Over a period of six months, the research will employ both qualitative and quantitative methodologies to provide a comprehensive analysis of contract termination dynamics.

The significance of this study lies in its potential to contribute to the existing literature on construction contract termination. By identifying the causes and effects of termination, the research will generate valuable insights for policymakers and practitioners, offering effective strategies to manage and mitigate the consequences of contract termination. Public construction projects play a vital role in delivering social and economic benefits, and the implications of their premature termination can be substantial. By conducting this research, policymakers and practitioners will gain a better understanding of contract termination dynamics and be equipped with knowledge to optimize project management practices and minimize the negative impacts associated with termination.

To accomplish the research objectives, a budget of 20,000 ETB has been estimated. This allocation will facilitate rigorous data collection and analysis, ensuring the research provides reliable and insightful findings. Ultimately, the findings of this study will contribute to the existing body of knowledge on construction contract termination and serve as a valuable resource for stakeholders involved in public construction projects.

Key words: - Public construction project, contract termination, cause-effects

1. INTRODUCTION

1.1. Background of the Study

Public construction projects play a vital role in developing a country's infrastructure and are often executed through contracts between the government and private companies. The termination of a contract can cause disagreements and lead to legal action, monetary damages for both parties, and postponement of the project's completion. The consequences of these problems are far-reaching and include significant economic losses, negative impacts on local communities, and reduced public trust in government institutions.

SNNPRS Industrial Parks Development Corporation main mission is to provide Industrial buildings with good quality of infrastructure that support the goal of Manufacturing industry growth. The corporation have been constructing industrial buildings with infrastructure continuously from 2009 E.C up to 2015 E.C. Through this year's many projects have been completed successfully but currently the challenge of contract termination is prominent issue.

Construction projects are known for their uniqueness and complexity and potential for disputes, and public construction projects are no exception. When a public construction contract is disputed or terminated, it can lead to different types of losses to all parties. Construction delays, cost overrun, poor quality are common characteristics of poorly managed project; when performance is not controlled and early mitigation measures are not taken; things go out of control hence claim, dispute arise which may or may not lead to contract termination. (Kebede and Tiewei, 2021)

The area of construction contract termination has received limited research attention due to various factors, including a lack of data, reluctance to take responsibility, negative perceptions of abandoned projects, poor project closure practices, and insufficient knowledge of construction contract law. However, it is important to recognize that valuable lessons can be learned from both successful and failed projects. While government offices tend to highlight their successful projects while ignoring failures, there is a need to shift this mindset and embrace the opportunity to learn from project failures in order to avoid repeating the same mistakes. While contract termination was not previously a common issue in public construction projects, the increasing occurrence of claims, disputes, and contract terminations necessitates a focus on mitigating the impact of terminations and minimizing their frequency.

1.2. Statement of the Problem

Public infrastructure projects play a crucial role in driving economic development, improving quality of life, and promoting sustainability. Industrial buildings managed by the SNNPRS Industrial Parks Development Corporation are currently experiencing challenges such as contract termination, litigation, poor project delivery, cost overruns, and budget constraints.

The construction industry in Ethiopia is plagued by challenges including cost overruns, poor performance, and construction delays. Studies have consistently highlighted the need for improved project planning and design, enhanced stakeholder communication and collaboration, stronger project supervision, reduced corruption, adequate funding, and enhanced workforce capacity. Gadisa and Zhou (2021) identified factors such as lack of skilled labor, poor planning and design, inadequate project supervision, and corruption as major causes of poor performance in public construction projects in Ethiopia. Koshe and Jha (2016) identified design changes, material shortages, payment delays, site management issues, and inadequate contractor experience as primary causes of construction delays. Ayalew, Dakhli, and Lafhaj (2016) emphasized the unsatisfactory level of construction project management practices, particularly in terms of safety, risk, and time management, and proposed solutions such as adopting modern project management approaches and enhancing stakeholder collaboration.

According to my desk-study and observation public construction project delay, cost overrun claim and dispute are common characteristics of the industry but in SNNPRS industrial parks development corporation currently the experience of contract termination and litigations becoming the prominent challenge.

Despite the importance of the issue of construction contract termination, In Ethiopia there is a lack of database on rate of contract termination of public construction projects; and researches that explores the causes and effects of public construction contract termination are rare. Therefore, this research aims to fill this gap by identifying the causes of contract termination, evaluating its effects, developing effective mitigation strategies,

1.3. Significance of the study

Public construction projects are major provider for infrastructure requirement of public service like hospitals, schools, Industrial buildings and transportation, water supply etc. Social and economic Benefits of public construction projects are achieved after completion of projects or else it is a waste of scarce resource of capital. Most development Goals and poverty alleviation projects and programs are highly linked public construction projects.

Termination of public construction projects can hinder the goal of public construction projects and can have many implications. This study tries to address the issue of causes and effects of contract termination on different stakeholders.

1.3.1. Clients (Project Owners):

First Understanding the causes of contract termination helps clients identify risks and develop strategies to mitigate them, ensuring better cost control and budget adherence. Secondly Insights into termination causes enable clients to make informed decisions during procurement, select capable contractors, and establish effective contract provisions for timely project delivery. And finally, the significance of Studying termination factors helps address communication gaps and conflicts, promoting better stakeholder management and collaboration.

1.3.2. Consultants (Designers, Engineers, Architects):

Identifying design-related issues contributing to termination allows consultants to optimize designs, reducing the probability of contract termination. The second important point for Studying termination helps consultants improve their contract administration practices, ensuring smoother project execution and reducing termination risks.

1.3.3. Contractors:

Studying termination factors allows contractors to assess risks and develop proactive risk management strategies, improving project planning and communication. And Analyzing termination reasons helps contractors identify areas for improvement, enhancing project management practices and resource allocation.

1.3.4. Users of Public Construction Projects:

Understanding termination factors helps users manage expectations and anticipate potential delays, ensuring uninterrupted access to public services. Studying termination factors enables users to demand high-quality construction projects, ensuring their needs are met effectively.

Overall, studying contract termination benefits all stakeholders by enabling better cost control, project delivery, stakeholder management, risk mitigation, performance improvement, service delivery, and quality assurance. It facilitates informed decision-making and promotes successful project outcomes.

1.4. Research Questions

1. what is the extent contract termination in the case of SNNPRS IPDC?
2. What are the major causes of contract termination in (SNNPRS IPDC)?
3. What are the effects of contract terminated construction projects in (SNNPRS IPDC)?
4. What are the appropriate mitigation measures to minimize PIBIP termination?

1.5. General Objective of the Study

The general objective of this study is to assess the main causes and effects of construction contract termination on public building projects in the SNNPRS (IPDC).

1.5.1. Specific objectives of the Study

1. To measure the extent of contract termination in PIBIP.
2. To identify causes of contract termination in PIBIP.
3. To evaluate the effects of contract termination on PIBIP.
4. To develop effective mitigation strategies for the PIBIP contract termination.

1.6. Scope of the study

This study will focus on public Industrial buildings and infrastructure construction projects owned by the SNNPRS IPDC. The projects are Located in various zonal cities. The regional construction bureau, zonal urban development and construction department offices, and some private consultants are the main providers of consulting service.

The study will collect and analyze primary and secondary data from projects that have been done from 2009 until 2015 in E.C so as to study the causes and effects of contract termination.

2. Literature review

2.1. Definition of Terms

Some important definitions by (Murdoch, J., and Hughes, W. (2015). Construction Contracts: Law and Management. Taylor and Francis.)

2.1.1. Construction Contract:

A construction contract is a legally binding agreement between a client and a contractor for the delivery of construction work, which typically includes building or civil engineering projects. The contract outlines the scope of the work, the timeline for completion, and the price to be paid for the work, as well as any other relevant terms and conditions. The contract may be based on a standard form or may be customized to suit the specific needs of the project.

2.1.2. Contract Termination:

Contract termination refers to the ending of a construction contract by either party before the work has been completed. Termination may be for convenience, which allows a party to end the contract for reasons other than a breach by the other party, or for cause, which allows a party to end the contract due to a specific breach by the other party. The grounds for invoking a termination clause will depend on the specific terms of the contract and the circumstances surrounding the termination. The termination of a contract may have significant implications for both parties, including financial losses, reputational damage, and delays to the project. Therefore, it is important to carefully consider the reasons for termination and to follow the appropriate procedures as outlined in the contract.

2.1.3. Types of Termination

Contract termination can occur under different circumstances. **Termination for convenience** allows a party to end a contract for reasons unrelated to a breach, such as changes in market conditions or budget constraints. On the other hand, **termination for cause** enables contract termination due to a specific breach by the other party, including failure to meet deadlines or deliver goods/services. **Termination for default** is similar to termination for cause but specifically applies to situations where one party fails to fulfill their contractual obligations. Each type of termination has its own criteria and procedures outlined in the contract, ensuring that the termination is carried out appropriately and in accordance with the terms agreed upon.

2.1.4. Grounds for Termination

Contract termination can occur due to various reasons. *Non-performance* arises when one party fails to fulfill their contractual obligations, such as late delivery or inadequate work. *Insolvency* of a party may also lead to termination, although the grounds for termination in such cases depend on the specific circumstances and contract terms. *Breach of contract* occurs when one party fails to meet their contractual obligations, and the grounds for termination will vary based on the contract terms. Additionally, *force majeure* events, including natural disasters or acts of war, can make it impossible or impractical to continue with the contract, potentially resulting in termination. Each of these situations presents distinct grounds for termination, and the specific criteria for termination will depend on the contractual provisions and the unique circumstances surrounding the contract.

2.2. Causes of Contract Termination in Public Construction Projects

According to the research article by Riveros et al. (2022), and Shinkafi (2021) several critical factors influence early contract termination in public design-build projects in developing and emerging economies. The following are some of the causes of construction termination highlighted in the study:

Early contract termination in public design-build projects can occur due to a variety of factors. One common cause is delays in project execution, which can be attributed to inadequate project planning. For example, if a construction project in Ethiopia lacks a comprehensive and well-structured plan, it can lead to delays and subsequent termination. Poor project management practices, such as ineffective coordination and resource allocation, can also contribute to project delays and early contract termination. Another factor is the shortage of skilled labor, where the lack of qualified workers hampers project progress and may result in contract termination. For instance, if a construction project in a developing economy like Ethiopia struggles to find skilled workers in a timely manner, the contractor may face difficulties in meeting project milestones, leading to termination.

Financial difficulties are another significant cause of early contract termination. When public design-build projects face budget constraints or insufficient funding, it becomes challenging for contractors to proceed with the project. For example, if a public construction project in Ethiopia experiences a funding shortfall due to changes in government priorities or economic constraints, the contractor may be forced to terminate the contract due to the inability to secure necessary resources. Disputes and conflicts between contracting parties can also result in early termination. These conflicts can arise from various reasons, such as disagreements over project scope, quality of work, changes in project design, or payment issues. If the parties fail to resolve these disputes amicably, it may lead to contract termination.

Political instability is another factor that can contribute to early contract termination. In developing and emerging economies, changes in government policies and regulations can create uncertainty and disrupt ongoing projects. For instance, if a construction project in a politically unstable country experiences a sudden shift in government priorities or regulations, it can result in contract termination as the project is no longer aligned with the new policies. Inadequate legal frameworks and regulations can also pose challenges. When there are ambiguous or unclear legal guidelines and procedures for dispute resolution, it becomes difficult to resolve conflicts, increasing the likelihood of early contract termination.

In the context of Nigeria, project abandonment is often driven by inadequate funding. For instance, if a building project in Nigeria experiences a lack of funds or mismanagement of funds, it can lead to abandonment as the contractor cannot sustain the project without sufficient financial resources. Poor project planning is another contributing factor, where inadequate feasibility studies, risk assessment, and lack of proper project management strategies can result in project failure and subsequent abandonment. Political interference and bureaucracy can further exacerbate project abandonment issues. Changes in government policies, bureaucratic delays, and excessive red tape can hinder project progress and lead to abandonment. Insecurity, such as terrorism and banditry, is also a significant concern in Nigeria, causing project abandonment in areas prone to such insecurity due to safety concerns. Lastly, the absence of technical expertise and skilled labor can undermine project success and lead to abandonment. If a construction project lacks qualified professionals and proper training, it can compromise the quality of work and result in project failure, ultimately leading to abandonment.

2.3. Effects of Contract Termination on Public Construction Projects

Project termination can have negative consequences for both the contractor and the client. The potential consequences and available damages depend on the reasons for termination and the terms of the construction contract. According to Surahyo, A., 2018.and (Riveros et al., 2022) contract termination can have significant negative impacts on both parties involved, including project delays, financial losses, reputational damage, legal disputes, decreased morale, and discouragement of future investment.

The termination of a public construction project can have various effects on both parties involved. One significant consequence is the occurrence of cost overruns and delays, which lead to additional expenses. The contractor may seek damages for the work already completed, while the client may incur costs in completing the project with a new contractor. This can result in financial losses for both parties.

Furthermore, project termination often triggers legal disputes and possible litigation, particularly when it is a result of a breach of contract. Damages sought by the non-breaching party may include direct costs associated with the breach, such as repairing defective work, as well as indirect costs like lost profits. Liquidated damages clauses, commonly found in construction contracts, specify the amount of damages to be awarded if the project is terminated prematurely.

These damages aim to compensate the non-breaching party for the costs incurred due to the delay or termination. Additionally, parties may seek compensation for lost profits if the project termination results in financial losses. In some cases, the prevailing party in a contract dispute may be entitled to attorney's fees, adding to the financial burdens for the losing party.

The effects of contract termination on public construction projects, therefore, encompass cost overruns, delays, legal disputes, financial losses, and the potential awarding of attorney's fees. These consequences highlight the importance of effective contract management, adherence to contractual obligations, and the implementation of mitigation strategies to minimize the likelihood of termination and its detrimental effects.

2.4. Mitigation Strategies for Public Construction Project Contract Termination

According to (solène rowan, 2011), Strategies and best practices to prevent project termination: Mitigation strategies play a crucial role in minimizing the risk of contract termination in public construction projects. Firstly, clear and detailed contract terms are essential to provide a solid foundation for the project. By ensuring that the contract outlines the scope of work, timelines, and expectations, both parties have a clear understanding of their obligations. It is also important to include provisions for change orders and dispute resolution, allowing for flexibility and a structured approach to resolving conflicts that may arise during the project.

Proper project planning is another vital mitigation strategy. Developing a comprehensive project plan that includes timelines, milestones, and goals helps establish a roadmap for successful project execution. Conducting thorough risk assessments allows for the identification of potential issues, enabling the development of contingency plans to address and mitigate those risks effectively.

To mitigate the risk of contract termination, adequate allocation of resources is crucial. Ensuring that the necessary resources, including materials, equipment, and skilled personnel, are available and allocated appropriately enhances the project's chances of success. Building a skilled and experienced team with a track record of successful project delivery further strengthens the project's foundation. Providing continuous training and development opportunities enables the team to enhance their skills and knowledge, reducing the likelihood of performance gaps and potential termination triggers.

Regular monitoring and evaluation are vital in mitigating the risk of contract termination. By monitoring project progress regularly and evaluating performance against established benchmarks, any issues or areas for improvement can be identified promptly. Taking corrective action in a timely manner ensures that problems are addressed proactively, helping to maintain project momentum and minimize the likelihood of termination. Additionally, fostering a collaborative environment and establishing effective conflict resolution mechanisms encourage parties to work together towards project success, preventing disputes from escalating to a point where termination becomes the only option.

In summary, mitigation strategies such as clear contract terms, proper project planning, allocation of adequate resources, building a skilled team, regular monitoring and evaluation, and effective collaboration and conflict resolution mechanisms are key to reducing the risk of contract termination in public construction projects. By implementing these strategies, stakeholders can proactively address challenges, minimize disruptions, and enhance the likelihood of project success and completion.

2.5. Summary of reviewed articles

Gadisa and Zhou (2021) utilize structural equation modeling to explore influential factors contributing to poor performance in public construction projects in Ethiopia. The study emphasizes the significance of inadequate planning, financing issues, and lack of skilled labor on project performance. Koshe and Jha (2016) investigate causes of construction delays in the Ethiopian construction industry, identifying factors such as poor project planning, inadequate site management, and delays in material procurement. Ayalew, Dakhli, and Lafhaj (2016) assess the performance and challenges of the construction industry in Ethiopia, highlighting issues such as inadequate project management practices, limited access to finance, and inadequate government support. Ade-ojo (2017) examines the implications of construction contract termination for infrastructure development in Ekiti State, Nigeria, emphasizing the need for effective dispute resolution mechanisms.

Additionally, studies such as Riveros et al. (2022) and Shinkafi (2021) investigate early contract termination and project abandonment, providing insights into critical factors and potential solutions in developing and emerging economies. The researches by Murdoch and Hughes (2007), Surahyo (2018), and Solène Rowan (2011) provide comprehensive understandings of construction contracts, their legal aspects, and remedies for breach of contract. Al-Zwainy, Mohammed, and Varouqa (2018) utilize root cause analysis to diagnose the causes of failure in the construction sector, emphasizing the importance of addressing these causes. Prince Elisha Nsiah-Asamoah (2019) assesses the causes, effects, and solutions to abandoned building projects in Ghana, highlighting factors such as inadequate funding and poor project management. Kebede and Tiewei (2021) explore the relationship between public work contract laws, project delivery systems, and project efficiency in Ethiopia, providing insights for improving contract effectiveness.

2.6. Identified Research gaps in construction termination:

After a reviewing literature in relation to public construction project termination the main research gaps identified are as follows

There are several research gaps within the field of construction, particularly in the context of Ethiopia and other developing economies. Firstly, there is a limited understanding of influential factors that contribute to poor performance and delays in construction projects beyond commonly identified factors such as inadequate planning, financing issues, and lack of skilled labor. Secondly, there is a need for more comprehensive research on contract termination and project abandonment, specifically examining the causes, effects, and potential solutions in these areas. Thirdly, the relationship between contract laws, project delivery systems, and project efficiency requires further investigation to better understand their impact. Additionally, there is a research gap in exploring and comparing remedies for breach of contract, particularly in terms of protecting performance. Lastly, there is a need for more focused research on the specific causes, effects, and solutions related to abandoned building projects, particularly in developing countries like Ethiopia. Addressing these research gaps will contribute to a deeper understanding of the challenges in the construction industry and enable the development of effective strategies for improving project performance and mitigating issues such as delays, contract termination, and project abandonment.

3. MATERIALS AND METHODS

3.1. Introduction

This chapter mainly explains how the study was conducted, the applied methods and techniques in data collection and the reasons as to why they were used according to the research aims and main objectives of the study. The chapter involves discussion of the research process, the selection of the study area, sampling methods and justification and sources of data used in the study. Analytical techniques used in analyzing the data for the study are also discussed.

The study adopted a mixed approach. The views from Contractors, Consultants and Client, working on construction of public projects Industrial buildings and infrastructures for the industrial building's accessibility and functionality in SNNPRS region will be collected via a questionnaire survey and a case study will be done on selected projects by systematic random sampling method. The questionnaire will be designed according to the objectives of research by reviewing literature dealing with public construction projects termination and related relevant topics. The sources of literature review included relevant books, case law, journals, magazines, dissertations and seminar proceedings. The review of literature provided useful information on the concept of public construction projects termination, the causes of construction contract termination, effects of construction contract termination, mitigating measures to of construction contract termination on the public building construction projects.

3.2. Description of the study area

The study area of this research is In SNNPRS Ethiopia which is located in the south western part of Ethiopia between 4 0 27' – 8 0 30' N latitude and 34 0 21' – 39 0 11' E longitude. The altitude of the region varies from 376 to 4207 masl (meters above sea level) and the yearly average rainfall ranges from 400-2200mm.

In the region limited number of public construction projects, including infrastructure development, such as road construction, building construction, and water supply projects are under-going construction; the study will focus on Industrial building and infrastructure projects that is built and owned by SNNPRS Industrial Parks Development Corporation. The study area includes17 zonal cities that is worabe, butajirra, wolikite, hossana, halaba, yem, durame, sheneshecho, hadero, wolayita sodo, boditi, areka, arbaminch, swola, jinka, dilla, yirgachefe, konso (Segene gumayede)

Figure 3. 1: Map of the Study Area

Source https://upload.wikimedia.org/wikipedia/commons/2/21/Regions_of_Ethiopia_EN.svg

3.3. Study Design

A mixed approach was adopted in this research. The views from Contractors, Consultants and client, working on public industrial building and infrastructure construction projects in the region will be collected via a closed as well as open-ended questionnaire surveys in addition to the archival/desk study that is conducted to arrive at the research objectives set forth in this research. The questionnaire will be designed according to the objectives of research by reviewing different literatures dealing with contract termination and other relevant topics which enable the researcher to come up with good theoretical background in relation to the topic of study. The reviewing relevant literature included relevant books, case law, journal articles, magazines, on line web sources, dissertations and seminar proceedings

The questionnaire will be designed based on information identified as potential causes of terminated PIBIP, the effect of termination on public construction projects and remedies to the terminated public construction projects.

The list of project owners, consultants, and contractors participated in public industrial building construction projects in SNNPRS IPDC was collected from zonal urban development and construction department, SNNPR design and construction supervision agency and SNNPRS IPDC construction projects administering office. Building projects constructed since year 2007Ec up to 2015Ec were considered for the study. Finally, the data was analyzed and interpreted using both descriptive and analytical approach.

3.4. Study Population

The population in research setting refers to the measurement of interest or collection of all possible individuals or objects (Mason et al., 1997). Hence, population includes all the potential individuals whom the measurement is being taken from. Accordingly, the study population in this research included contractors, consultants and clients involved in the public industrial building and infrastructure construction projects in SNNPRS and the archival documents in the SNNPRS IPDC offices.

Project owners, Consultants and contractors were identified based on their involvement in execution of PIBIP in the Region. All Consultants and contractors were considered. The basis for this selection is that, contractors and consultants under these categories are easy to be located and they are also expected to have professionals who have better experience and exposure to the problem of construction project Termination. Accordingly, 25 Grade (1-5) contractors and 30 Grade (6-9) contractors ,2 reginal consultants 12 zonal consultants and 1 client were identified as stake holders in PIBIP in the region to execute 400 fixed price agreement contracts with low level project complexity and volume with project duration up to 3-6 months , 25 unit rate price contracts with medium project complexity and volume with project duration up to 6-9 months and finally 9 projects unit rate price contracts with high project complexity and volume of project with duration up to 12 month up to 2 years.

3.4.1   Selected projects for study

Cluster up grading project with allocated budget of birr 212,226,166.76) and with total population size of 444 projects with fixed price contract type out of which 81 projects are terminated and the reset are completed or on going so document analysis will be don on sample of terminated projects i.e., sample size will be 68 using sample size table will use comparative analysis for completed projects with terminated projects

Production shade and site work /Industrial Building/projects (allocated contractual budget of around 200 million) and with total population size of 25 and contract type is unit price out of which 7 are completed the reset 17 are terminated sample size will be 25 using sample size table will use comparative analysis for completed projects with terminated projects

RTC (Rural Transformation Centers) projects with /Industrial Building and infra structure and water works /projects with contractual Budget (878,811,413.37) and with total population size of 11 projects all are ongoing projects with different progress (water works 6 projects), infra structure, main gate and building 5 projects all are ongoing projects with different progress. Due to sample size smallness all 11 projects will be taken as sample

3.5. Sampling Technique and Sample Size

Sampling as the process of selecting representative units of a population for a study in a research investigation and a sample is a portion of population selected for observation and analysis (LoBiondo-Wood and Harber, 1998).

A combination of purposive and systematic random sampling techniques will be used to conduct this study. The research samples from clients, consultants and contractors will be selected randomly from the collected list mentioned in section 3.3 above and then, the respondents from within the selected samples were chosen purposively based on their knowledge in the area of study, their qualification and years of experience.

To determine sample size, the sample size Table from web of research Advisor posted in Researchadvisor.com that suggests the optimal sample size based on a given a population size, a specific margin of error and a desired confidence interval was used. This can help researchers avoid the formulas altogether (Research Advisor, 2006). For this research, a confidence level of 95% and margin of error equal to 9% was referred to decide the sample size out of total population.

3.5.1. Sample size of Contractors

As discussed above in section 3.4, for cluster up grading a total of 68 contractors were identified to involve in the execution of cluster up grading projects whose projects are terminated. Due to the reason that contractors with lower Grade (7-9) and terminated contractors are hard to find only document analysis will be done and the knowledge level of SME contractors are lower analyzing the document with questioner will be enough. for production shade and site work projects a total of 17 contractors were identified to involve in the execution of production shade and site work projects whose projects are terminated all will be taken as sample and for RTC Building and infra structure projects a total of 5 contractors were identified to involve in the execution of projects building and infra structure projects and for 5 projects water works one contractor is identified. Hence, the total population size is 23. Accordingly, by using Sample size Table of properly linked excel spread sheet from web of research Advisor, the sample size was read to be 24 (Please refer Appendix E at the end of this paper). Hence 23 contractor companies to be considered in the research were selected by A purposive sampling procedure will be used to select project managers, contractors, and government officials who have experience with PIBIP construction project contract termination. And two questionnaires were distributed for each of the selected contractor company's staffs (One for the project manager and the remaining one for either construction engineer, office engineer or site engineer whomever assigned at hierarchical position next to the project manager). As such, a total of **46** questionnaires were distributed to professionals working under contractors PIBCP in the region.

3.5.2. Sample size of consultants

With respect to consultants, as per the recommendation of the web of research advisor, for the population size of 12, all the candidates should be considered for the collection of the required data. Hence the sample size here is equal to the population size (i.e., 12). Two of the consulting offices involved in the public construction projects in the regional were governmental organizations namely regional and zonal design and construction offices and private consultant. Hence 2 respondents from each of these three offices were selected and for each of the remaining, two questionnaires were distributed per company (One to the resident Engineer and one to either office engineer or site supervisor). As such, a total of **24** questionnaires were distributed to consultant staffs working in PIBIP in the region.

3.5.3. Sample size of Clients

As discussed above in section 3.4, a total of 2 types clients were identified to own public building construction projects in the region and zone. Hence, the total population size of client was 14. Accordingly, by using Sample size Table from web of research Advisor, the sample size of client organization was read to be 14. These 14 clients were systematically selected out of the population and were considered for the collection of the required data. Two questionnaires were distributed for each of the selected client company's staffs (One for the head of the respective office and one for the vice head who is concerned with the follow-up of the proper execution of the construction projects). As such, a total of **28** questionnaires were distributed to client staffs.

3.5.4. Desk Study

For the purpose of the desk study, ten building projects constructed over the past 7 years were considered.

3.6. Data Collection

The study data will be collected via both primary and secondary sources. In order to achieve the pre-set research objectives, the researcher carried out the desk study (secondary data source) and the views of the stake holders in the construction industry namely, clients, consultants and contractors were collected via questionnaire (primary source). Depending on the research questions as well the research objectives, a complete data collection from the respondents was done in appreciable manner.

Contract termination, payment process, condition of contract applied the level of delayed payments with respect to the term of agreement in the public building construction projects in the zone were extracted using desk study.

These data were collected and used to systematically investigate the extent of terminated projects existing in PIBIP in the region there by answering the first objective. After a critical look at the research objectives and existing literature, a well-structured questionnaire will be prepared and administered to the various respondents by the researcher. All the questionnaires have both closed and opened ended questions to certify consistency of respondent feedback. For the reason that it is not totally possible to design all questions as closed-ended, some of the questions will be left open-ended, to acquire numerical data or to lobby some written comment. A five-point Likert scale of 1 to 5 will be employed so as to measure the strength of respondent 's view or opinion on extents, causes, effects and remedies of terminated public building construction projects in the region.

3.7. Data Analysis

Both descriptive and inferential statistics were employed in the data analysis. In the analysis the 'relative importance index (RII)' method was adopted to establish the degree of the causes, effects and remedies of delayed payment in public building construction projects in the region. Likert's scale of five ordinal measures of agreement towards each statement (1, 2, 3, 4 and 5) will be used to calculate the RII for each factor that was used to determine the relative ranking.

4. Timeline

Time line (2023GC)

Research Activities	April			May			June			July			august			sept			remark
Developing proposal	▓	▓	▓	▓	▓														
Preparing interview questionnaires						▓	▓	▓											
Locating and defining the target group							▓	▓	▓										
Literature review									▓	▓	▓								
Data collection											▓	▓	▓						
Data processing												▓	▓	▓					
Data analysis and Interpretation													▓	▓	▓				
First draft writing														▓	▓	▓			
Draft literature and Complete remaining chapters														▓	▓	▓			
Revise overall draft															▓	▓	▓		
Submit draft and defense																▓	▓		
Correction of the final draft																	▓	▓	
Submission of research paper																		▓	

Timeline

5. Budget

Item	Unit	quantity	Unit cost	Total
Photo copying	No of pages	200	3	600
printing	No of pages	600	4	2,400
Transportation	No of trips	20	300	6,000
Internet data fee	Data fee/megabytes	200,000 megabytes	0.015	3000
Miscellaneous costs	Lump sum		3000	3,000
Contingency	Lump sum		5000	5,000
Total				<u>20,000.OO</u>

Budget

6. Expected outcomes:

The study is expected to provide the following outcomes:

A comprehensive understanding of the legal, financial, and performance consequences of contract termination and identification of major causes of Termination in public construction projects.

Decision-making strategies that will assist in making informed decisions about contract termination in public construction projects.

A case study of how the decision-making tools can be applied to a specific public construction project in the SNNPRS Industrial parks development corporation.

References

1. Gadisa, B., Zhou, H., 2021. Exploring influential factors leading to the poor performance of public construction project in Ethiopia using structural equation modelling. ECAM 28, 1683–1712. https://doi.org/10.1108/ECAM-12-2019-0689

2. Koshe, W., Jha, K.N., 2016. Investigating Causes of Construction Delay in Ethiopian Construction Industries 1, 18–29. https://doi.org/10.11648/j.jccee.20160101.13

3. Ayalew, T., Dakhli, Z., Lafhaj, Z., 2016. Assessment on Performance and Challenges of Ethiopian Construction Industry 2, 01–11.

4. Ade-ojo, 2017. Implication of construction contract termination for infrastructural development in Ekiti State Local Government Areas,.pdf. FUTY Journal of the Environment 11, 32–40.

5. Murdoch, J., Hughes, W., 2007. Construction Contracts: Law and Management, Fourth Edition.

6. Claudia Riveros 1, Angie L. Ruiz 2 , Harrison A. Mesa 1,3,* and Jose A. Guevara 2, 2022. Critical Factors Influencing Early Contract Termination in Public Design–Build Projects in Developing and Emerging Economies. https://doi.org/10.3390/buildings 12050614

7. Shinkafi, A M, A.M., 2021. Review of Possible Solutions to the Causes and Effects of Project Abandonment using Engineering Approach and Terminologies for Sustainable Economic Development in Nigeria. Bakolori Journal of General Studies 12, 3717–3725.

8. Surahyo, A., 2018. Understanding Construction Contracts. Springer International Publishing, Cham. https://doi.org/10.1007/978-3-319-66685-3

9. SOLÈNE ROWAN, 2011. REMEDIES FOR BREACH OF CONTRACT. A COMPARATIVE ANALYSIS OF THE PROTECTION OF PERFORMANCE. England.

10. Al-Zwainy, F.M.S., Mohammed, I.A., Varouqa, I.F., 2018. Diagnosing the Causes of Failure in the Construction Sector Using Root Cause Analysis Technique. Journal of Engineering 2018, 1–12. https://doi.org/10.1155/2018/1804053

11. Prince Elisha Nsiah-Asamoah, 2019. Assessment of Causes, Effects and Solutions to Abandoned Building Projects in Ghana. PARIPEX, International Peer Reviewed and Refereed Journal with Indexed Journal Platforms 1, 111–123. https://doi.org/10.15373/22501991

12. Kebede, S.D., Tiewei, Z., 2021. Public work contract laws on project delivery systems and their nexus with project efficiency: evidence from Ethiopia. Heliyon 7, e06462. https://doi.org/10.1016/j.heliyon.2021.e06462

13. Gadisa, B., Zhou, H., 2021. Exploring influential factors leading to the poor performance of public construction project in Ethiopia using structural equation modelling. ECAM 28, 1683–1712. https://doi.org/10.1108/ECAM-12-2019-0689

14. Kelley, K., Clark, B., Brown, V. and Sitzia, J. 2003. Good practice in the condust and reporting of survey research. International Journal for Quality in Health Care, 15(3), pp. 261- 66.

15. LoBiondo-Wood, G. and Harber, J. 2002.Nursing research: Methods and Critical appraisal for Evidence-based practice. (5 Ed.). St. Louis: Mosby Elsevier